WHERE IS THE MOON?

Misty Carty, Ph.D.

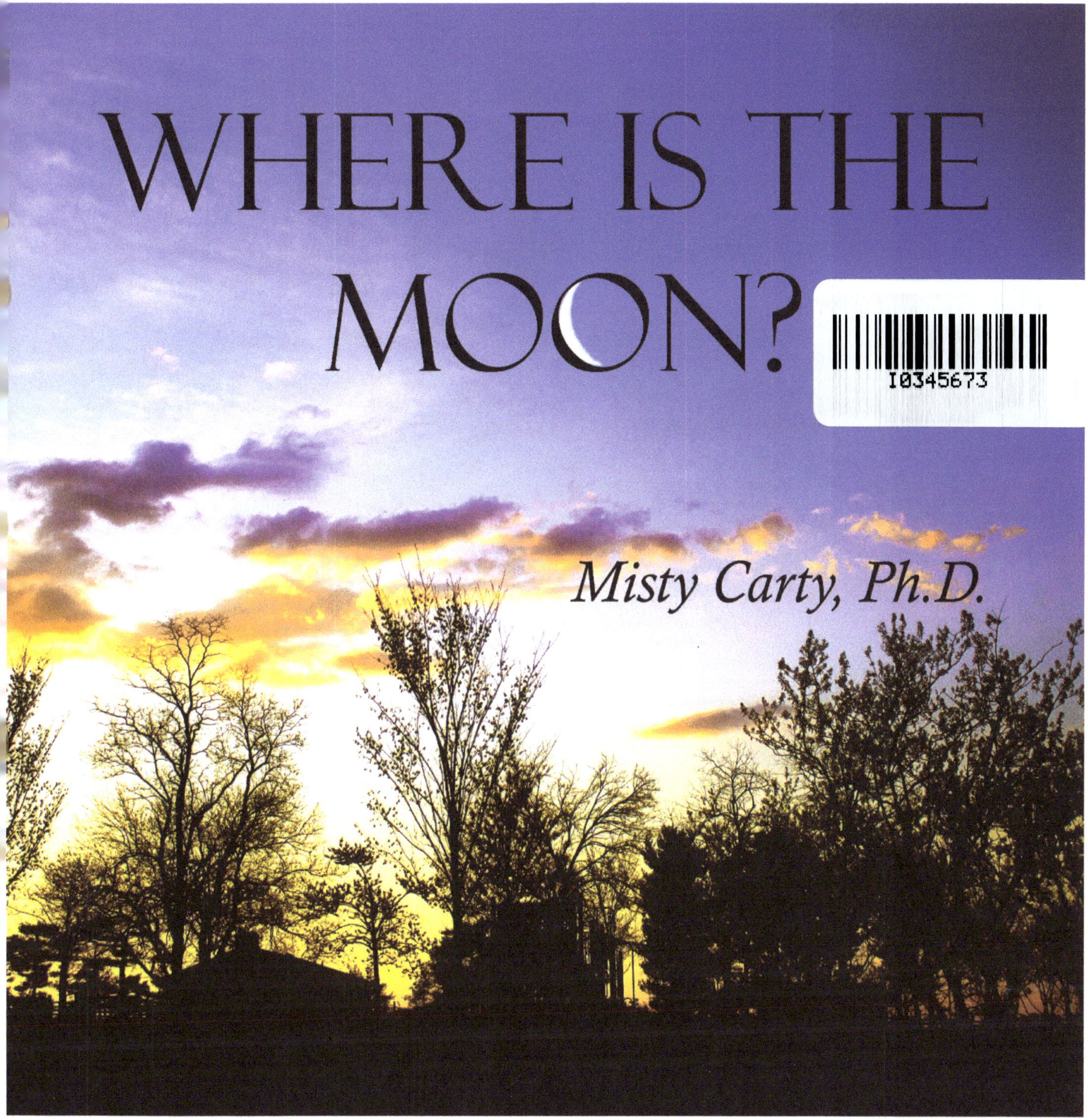

Every night before bed, I read my new favorite story, "Goodnight Moon". Then I say goodnight to the Moon from my window.

"Goodnight Moon!"

Oct, 4

"Goodnight Moon!"

Oct, 3

"Goodnight Moon!"

Oct, 2

"Goodnight Moon!"

Oct, 1

I see the Moon in a different spot each night!

"Goodnight..."

Wait, where is the Moon?

I cannot see the Moon from my window.
Mommy and I go outside to explore.

We look up and see the Moon!
It is much higher in the sky and I can see almost half of it!

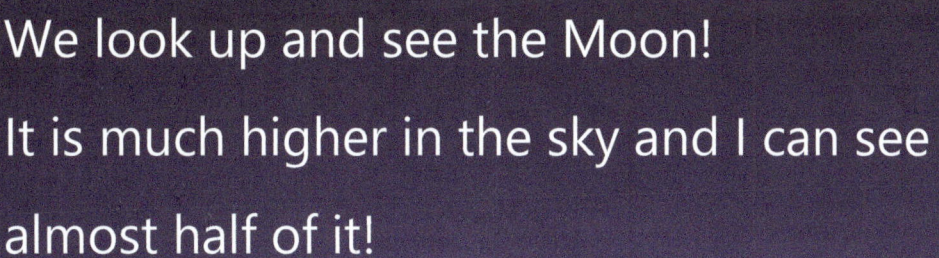

Oct, 5

"Goodnight Moon!" I say.

Now, each night before bed, I go outside with Mommy to say goodnight to the Moon.

I see more and more of the Moon each night.

For two nights it rains and I cannot see the Moon. When the rain stops, I go outside. But where is the Moon?

Mommy and I walk around our neighborhood.
And...there it is!

Oct, 12

"Goodnight Moon!" I say excitedly.

The next night, Mommy and I walk to say goodnight to the Moon again.

"Goodnight Moon!"

Oct, 13

Tonight, I see all of the Moon. It is barely above the trees. I wonder where I will see it tomorrow.

But, the next night the Moon is not there!

Mommy and I look all around, but we cannot find the Moon. I wonder where it is. Mommy says we will check for it tomorrow.

After I wake up, I go to my window to check for the Moon. I look out and see...

the Moon! I see the Moon in the morning!
"Good morning Moon!" I say happily.

Oct, 15

Now, each day I say good morning to the Moon instead of goodnight.

"Good morning..."

Wait, where is the Moon?!

I cannot see the Moon from my window again. Mommy and I go outside.

We look up and there it is!

"Good morning Moon!" I say.

Oct, 20

I see almost half of the Moon again. This time it is the left half.

Now, Mommy and I go outside to say good morning to the Moon.

For the next two days it is cloudy. When the sky clears I go out to look for the Moon.

But where is the Moon?

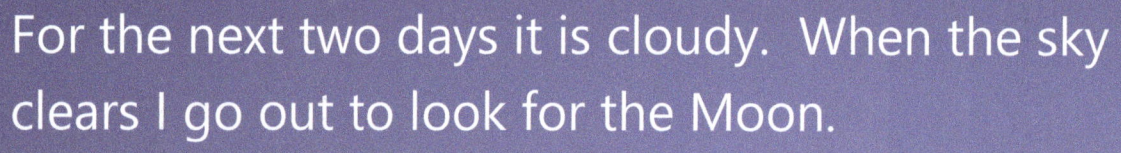

Mommy and I walk up the hill to the park and there it is!

Oct, 26

"Good morning Moon!" I say to my friend.

The next morning Mommy and I walk to the park.

"Good morning Moon!"

Oct, 27

The Moon is so thin now, like when I first saw it out my window.

The next morning, Mommy and I do not see the Moon. I wonder where the Moon is. Maybe it is not shining? Mommy says we will look for it and assures me we will see it again.

We look for the next couple of mornings and nights. But we do not see the Moon.

Then....

I see the Moon from my window before bed! It is up at night again.

"Goodnight Moon!" I exclaim.

Oct, 31

I see the Moon where I did a month ago.

I discovered how I see the Moon changes during a month.

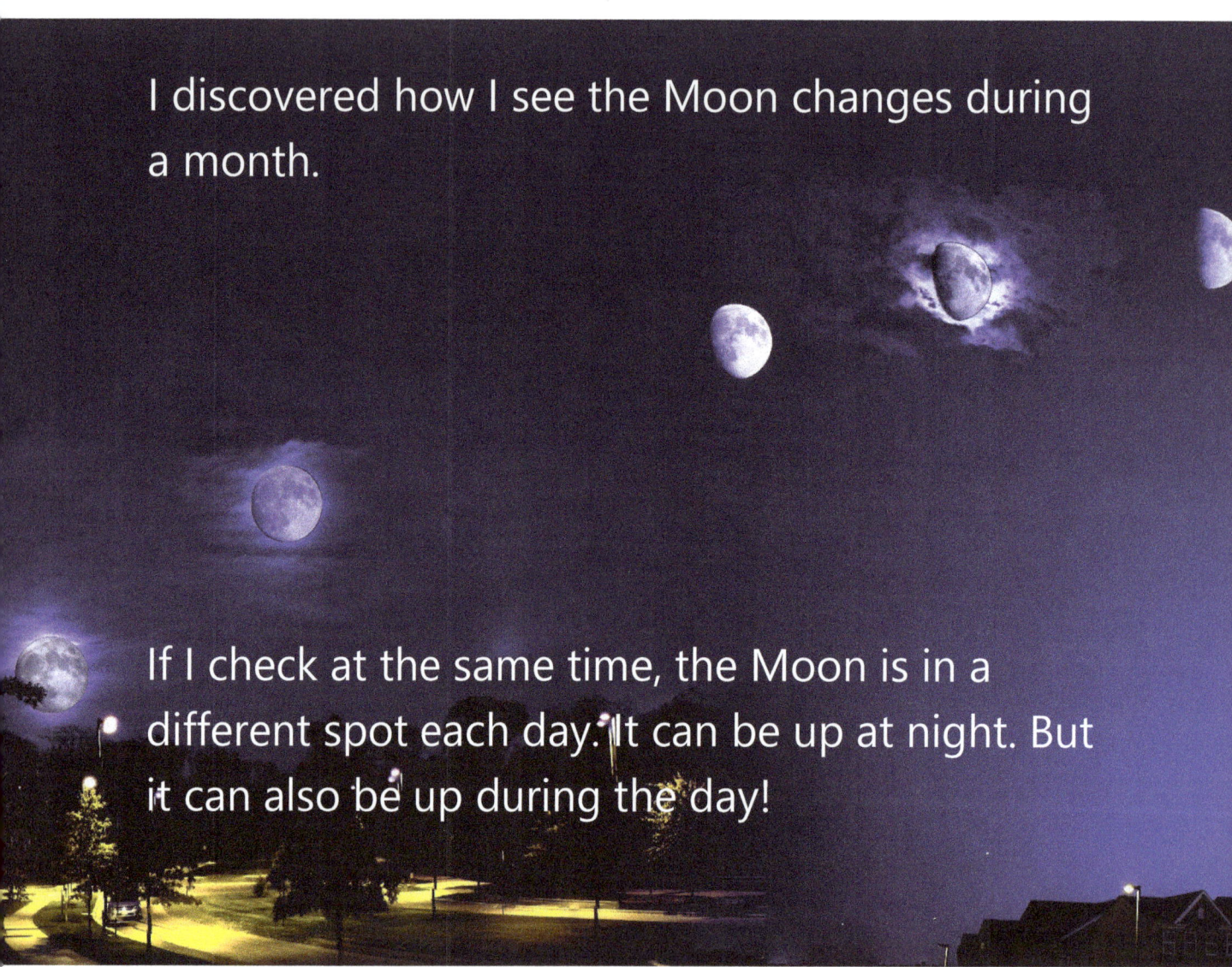

If I check at the same time, the Moon is in a different spot each day. It can be up at night. But it can also be up during the day!

The shape of the Moon also changes.

It starts as a thin sliver building to a full circle in the night sky.

Then the Moon lessens from a full circle back to a thin sliver in the morning sky.

Soon after, the Moon begins its cycle over again!

For Parents and Educators:

Introducing science to young children is fun! As any parent knows, kids are little scientists; experimenting and interacting with the world around them every day. Already curious, children love furthering their vocabulary and knowledge about the objects and actions they experience.

In this book, *Where is the Moon?*, your child will experience the fun of asking questions and finding answers! Your child is front and center determining where the Moon is each night, empowering them to make observations, ask questions about what they see, and - with a little help - find the answers. Throughout roughly a month, they will observe both the change in location of the Moon and its illumination. They'll see the waxing and waning of the Moon and see it drift eastward across the sky. After the Moon wains completely, they'll experience the joy of its return, discovering the variations they observe are actually a cycle!

As you read this book with your child, please use the following information to deepen your science experience.

Images of the Moon

The images of the Moon were taken each night and morning at the same time of day. This is to illustrate both the Moon's eastward progression across the sky and the variation of the Moon's illumination visible from Earth throughout a lunar cycle.

We first glimpse the Moon in the night sky just after New Moon. From Earth, only a sliver of the Moon is illuminated. As the lunar cycle progresses, we see a little more of the moon each night, approximately 7% more each night. We also see the Moon progress eastward across the sky, approximately 13° further east each night.

After Full Moon, the Moon is now visible in the morning sky! We see a little less of the moon each morning as it again progresses eastward.   As it approaches New Moon again, we see only a tiny sliver of the left side of the Moon. Then is reappears again just after New Moon where we originally saw it in the evening sky.

The changes we see in the Moon throughout a cycle are due to how it orbits the Earth.

Taking it Further

Observing the changes in the lunar cycle is something everyone can do! You can start at any time during a lunar cycle – beginning at New or Full Moon can be a fun way to begin your experiment. Pick a time of day that you will observe. It is easiest to see the first half of the lunar cycle – New to Full Moon- in the evening. The last half of the lunar cycle – Full back to New Moon – is easiest to see in the morning. Pick a place to observe with a clear view of the horizon. Make your observations! Note the position of the Moon on the sky and how much of the Moon you see. Collect your observations over roughly a month and discover where the Moon is throughout a lunar cycle!